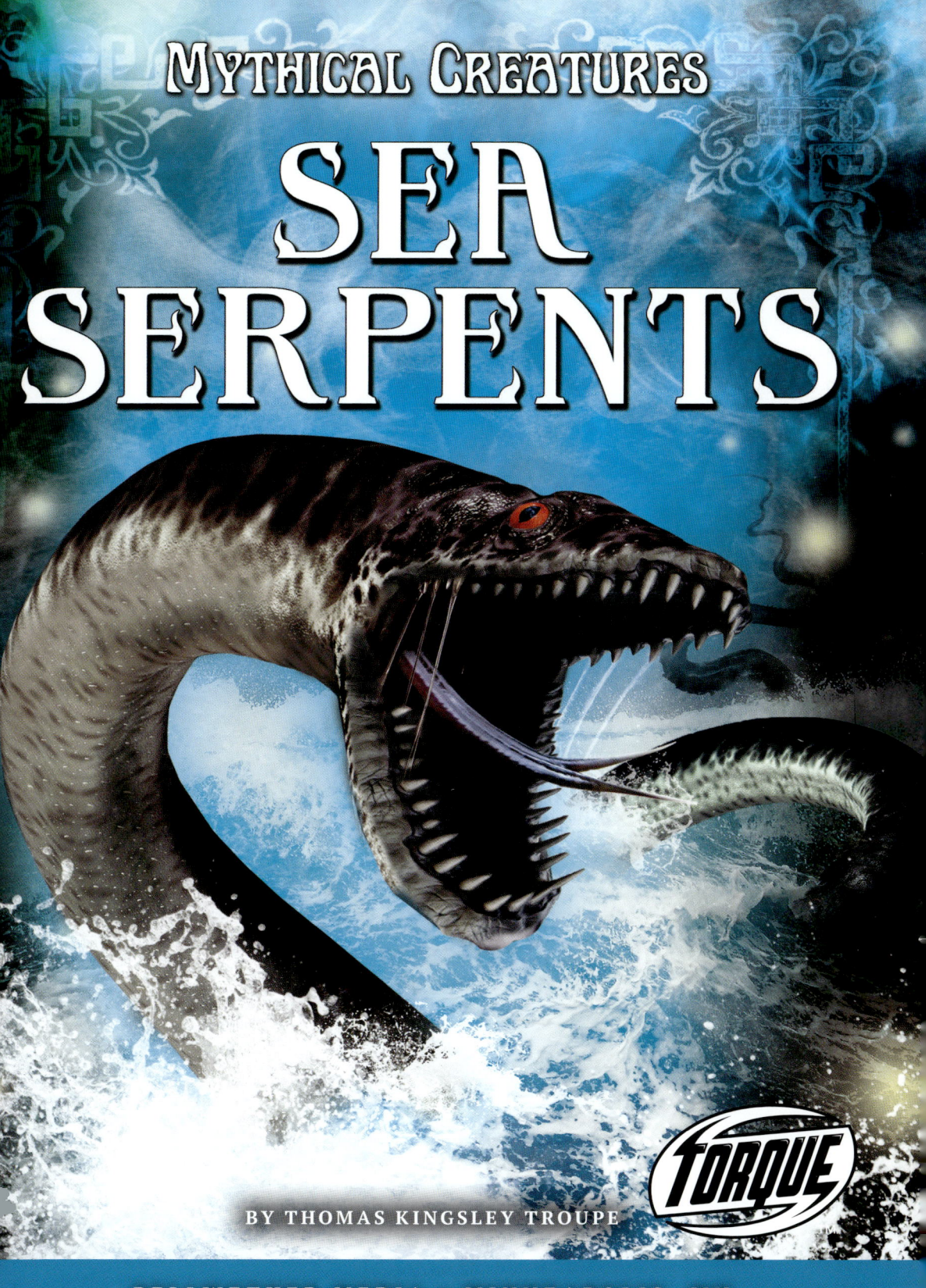

Torque brims with excitement perfect for thrill-seekers of all kinds. Discover daring survival skills, explore uncharted worlds, and marvel at mighty engines and extreme sports. In *Torque* books, anything can happen. Are you ready?

This edition first published in 2021 by Bellwether Media, Inc.

No part of this publication may be reproduced in whole or in part without written permission of the publisher.
For information regarding permission, write to Bellwether Media, Inc., Attention: Permissions Department, 6012 Blue Circle Drive, Minnetonka, MN 55343.

Library of Congress Cataloging-in-Publication Data

Names: Troupe, Thomas Kingsley, author.
Title: Sea serpents / by Thomas Kingsley Troupe.
Description: Minneapolis, MN : Bellwether Media, 2021. | Series: Torque : mythical creatures | Includes bibliographical references and index. | Audience: Ages 7-12 | Audience: Grades 4-6 | Summary: "Amazing images accompany engaging information about sea serpents. The combination of high-interest subject matter and light text is intended for students in grades 3 through 7"– Provided by publisher.
Identifiers: LCCN 2020046871 (print) | LCCN 2020046872 (ebook) | ISBN 9781644874660 (library binding) | ISBN 9781648341434 (ebook)
Subjects: LCSH: Sea monsters–Juvenile literature.
Classification: LCC QL89.2.S4 T76 2021 (print) | LCC QL89.2.S4 (ebook) | DDC 001.944–dc23
LC record available at https://lccn.loc.gov/2020046871
LC ebook record available at https://lccn.loc.gov/2020046872

Text copyright © 2021 by Bellwether Media, Inc. TORQUE and associated logos are trademarks and/or registered trademarks of Bellwether Media, Inc.

Editor: Rebecca Sabelko   Designer: Josh Brink

Printed in the United States of America, North Mankato, MN.

TABLE OF CONTENTS

A SWIMMING SERPENT	4
SERPENTS AROUND THE WORLD	10
THE SEA SERPENT TODAY	18
GLOSSARY	22
TO LEARN MORE	23
INDEX	24

A SWIMMING SERPENT

A huge sea serpent speeds through the water. Its body creates giant waves. The water crashes against your boat. It floods the deck.

You hold on tight as the serpent breaks through the water's surface. You look up. The monster stares at you and hisses. It is ready to strike!

Sea serpents have been a part of **mythology** for centuries. Some of the oldest stories of sea serpents are **creation myths**. Others come from sailors and their fear of the ocean.

Many sea serpents look like giant snakes with dragon heads. Their bodies are covered with scales. Some have horns on their heads. Giant fangs stick out of their mouths.

Chinese Dragons

Chinese dragons look a lot like snakes. They have long bodies and short legs. It is believed they control water.

Many people believed sea serpents lurked in the deepest ocean waters. The creatures would wait for ships to pass overhead. Then they attacked! They wrapped their bodies around boats to crush them.

Some believed sea serpents caused rough seas and giant waves!

Monster Maps
Some old maps of the seas included images of sea serpents. Sailors believed the maps helped them avoid danger.

Serpent Myths Around the World

Ogopogo
(Canada)

Ryujin
(Japan)

Awanyu
Tewa
(United States)

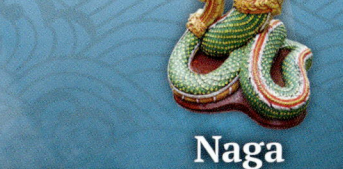
Naga
(India, Nepal, Indonesia, China, Cambodia, Malaysia, Thailand)

SERPENTS AROUND THE WORLD

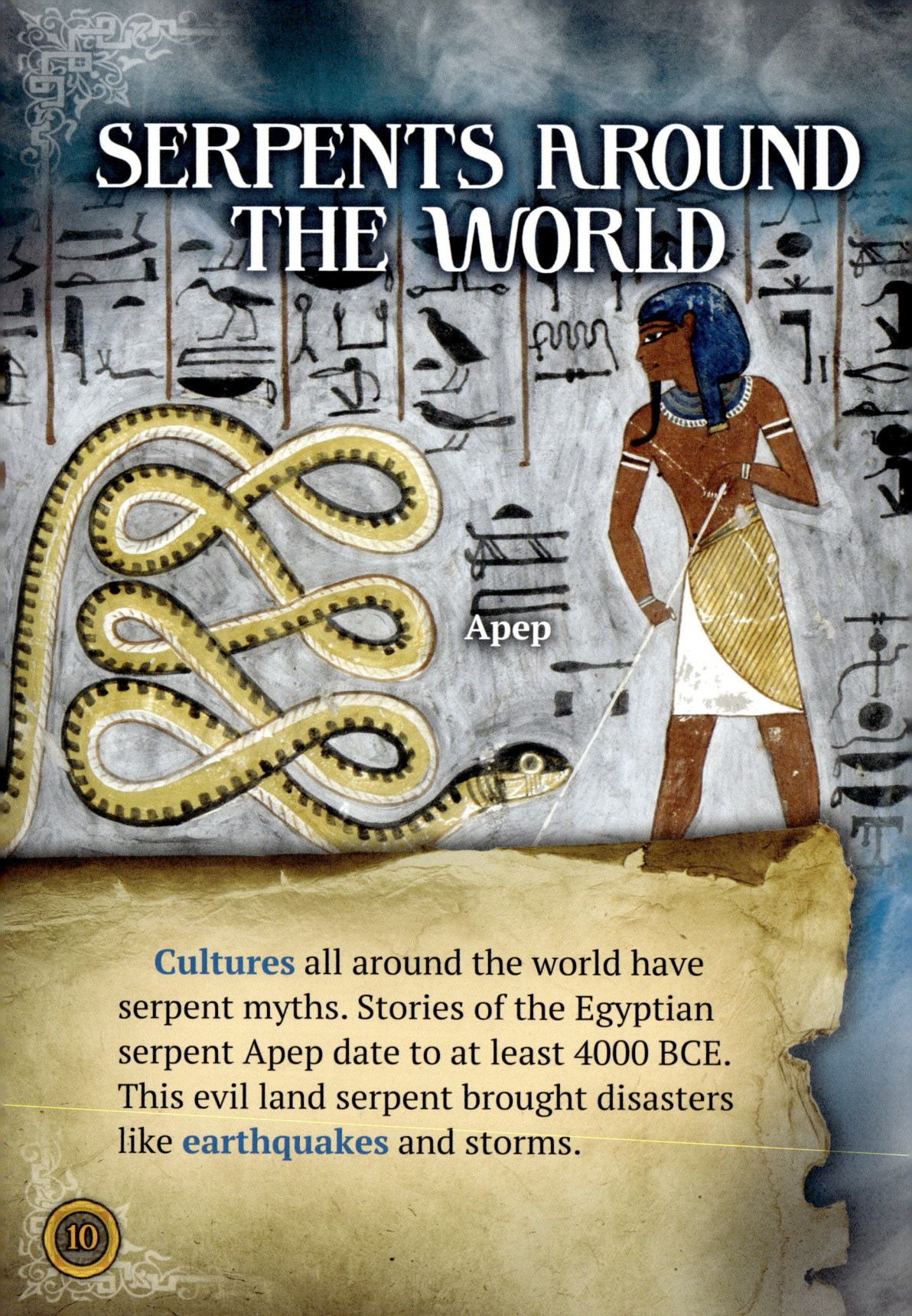

Apep

Cultures all around the world have serpent myths. Stories of the Egyptian serpent Apep date to at least 4000 BCE. This evil land serpent brought disasters like earthquakes and storms.

Mythical sea serpents first appeared in **Mesopotamia**. The goddess Tiamat was a **symbol** of **chaos**.

Around 350 BCE in Greece, Aristotle described a sea serpent that sunk a large ship. It also ate oxen.

Sea Serpent Origin

Mesopotamia =

A sea serpent is an important symbol in the **Hebrew Bible**. Leviathan is a giant monster. In some stories, it stands for evil. In others, it is a symbol for God's power to create.

Leviathan has many heads. Anyone who approaches the monster's mouths will die. Its jaws can bite through iron. Only God can **tame** the snake.

Leviathan

Sea Serpent Timeline

around 1750 BCE: The sea serpent Tiamat appears in a poem, but there are likely earlier mentions of the goddess

beginning around 1200 BCE: The Hebrew Bible includes a story about the sea serpent Leviathan

1846 CE: A sea creature is reported in the Chesapeake Bay and eventually named Chessie

Norse mythology included a fearsome sea serpent, too. Jormungand could wrap itself around the world. The serpent could swallow a god or a giant whole.

The thunder god Thor fought the mighty serpent. He defeated the beast but died from its **venomous** bite.

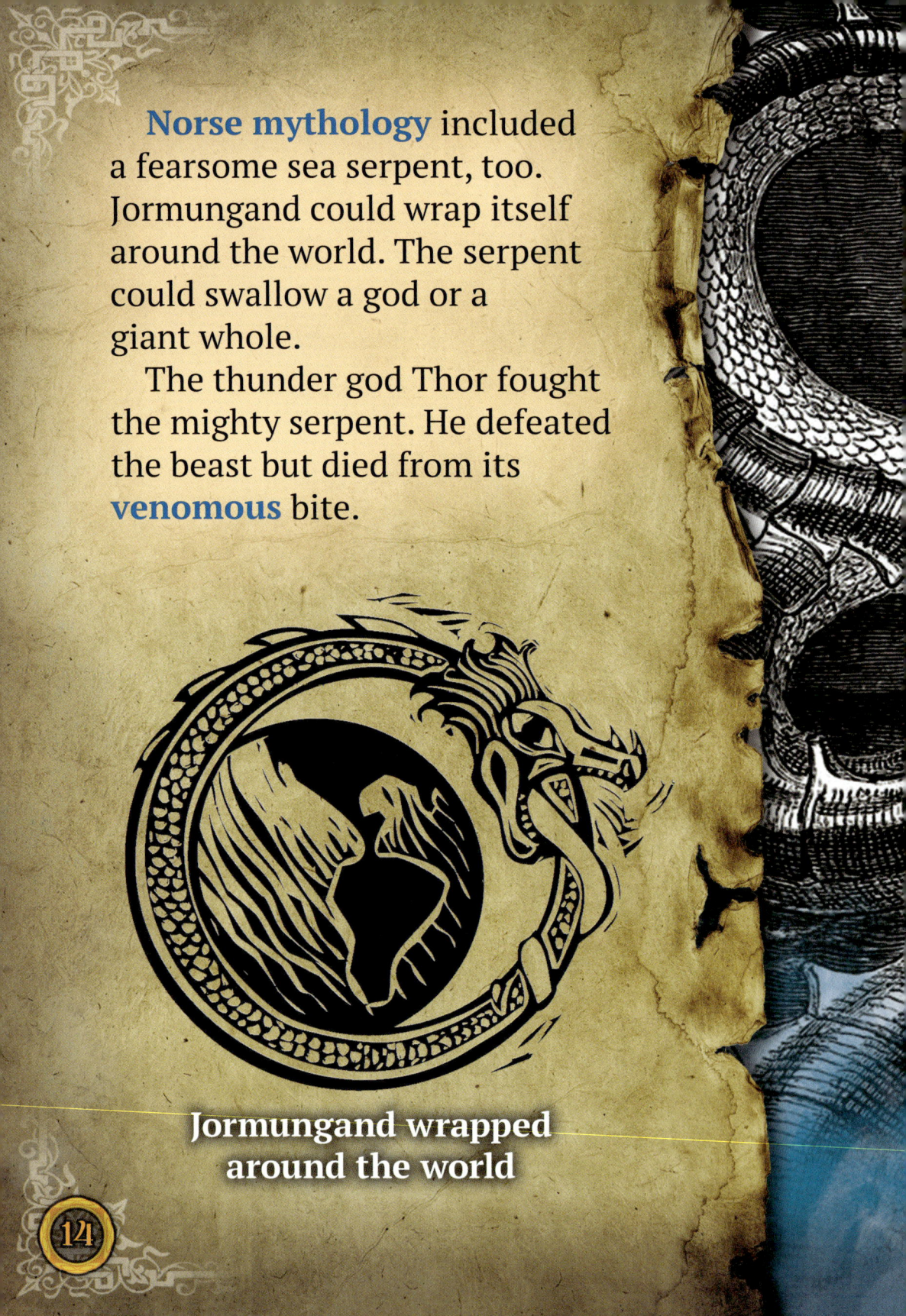

Jormungand wrapped around the world

Thor battling Jormungand

Sea serpent sightings have occurred for hundreds of years. But there are many explanations that prove these mythical monsters are not real.

Large collections of seaweed can look like a sea snake among the waves. The noses of basking sharks can also fool untrained eyes. From a distance, ribbonfish, oarfish, and sea lions may be mistaken for sea serpents, too.

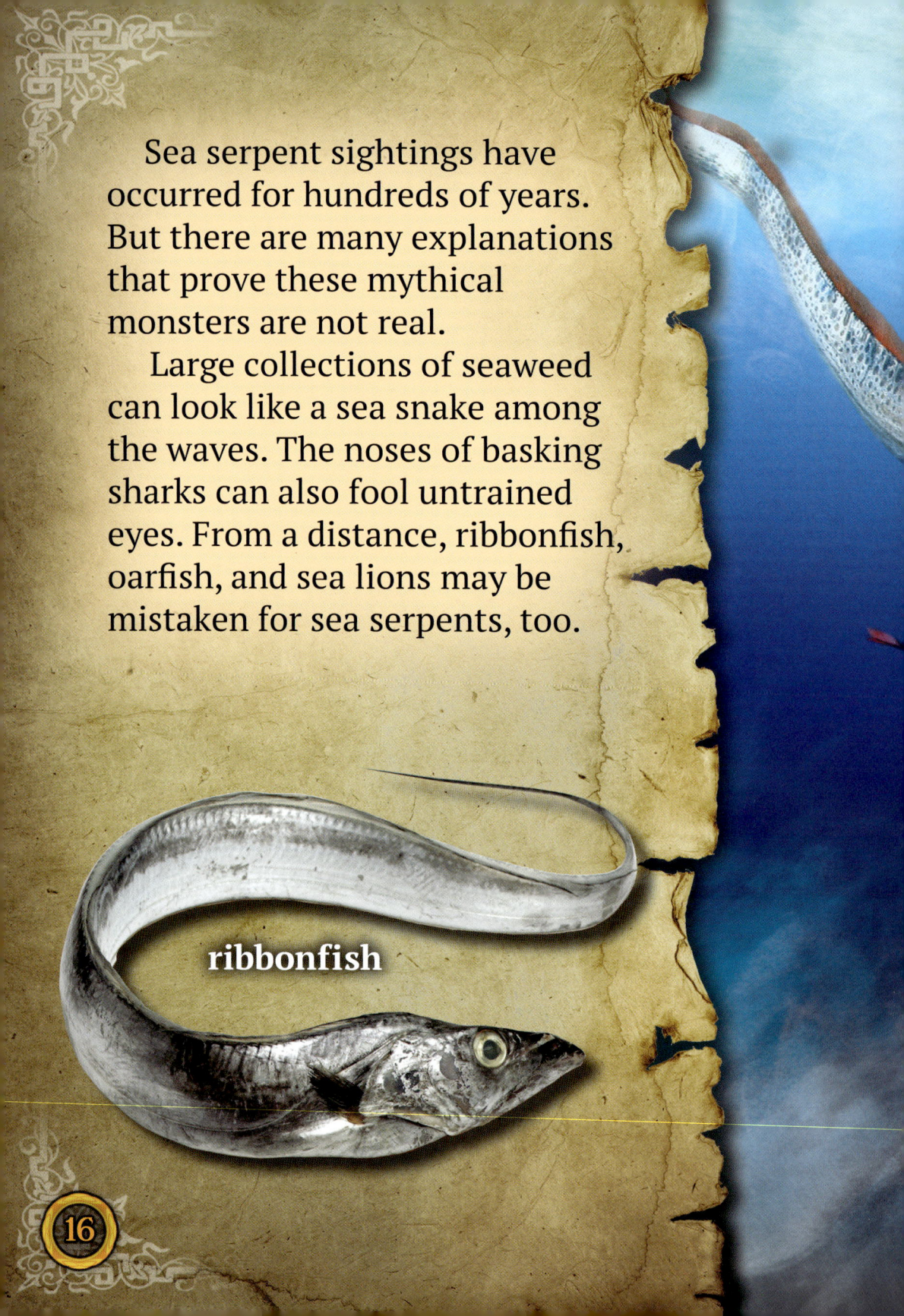

ribbonfish

oarfish

Chessie

Reports of a sea creature living in the Chesapeake Bay began in 1846. The bay is located off the coast of the United States. For nearly 150 years, people reported sightings. The creature was named Chessie.

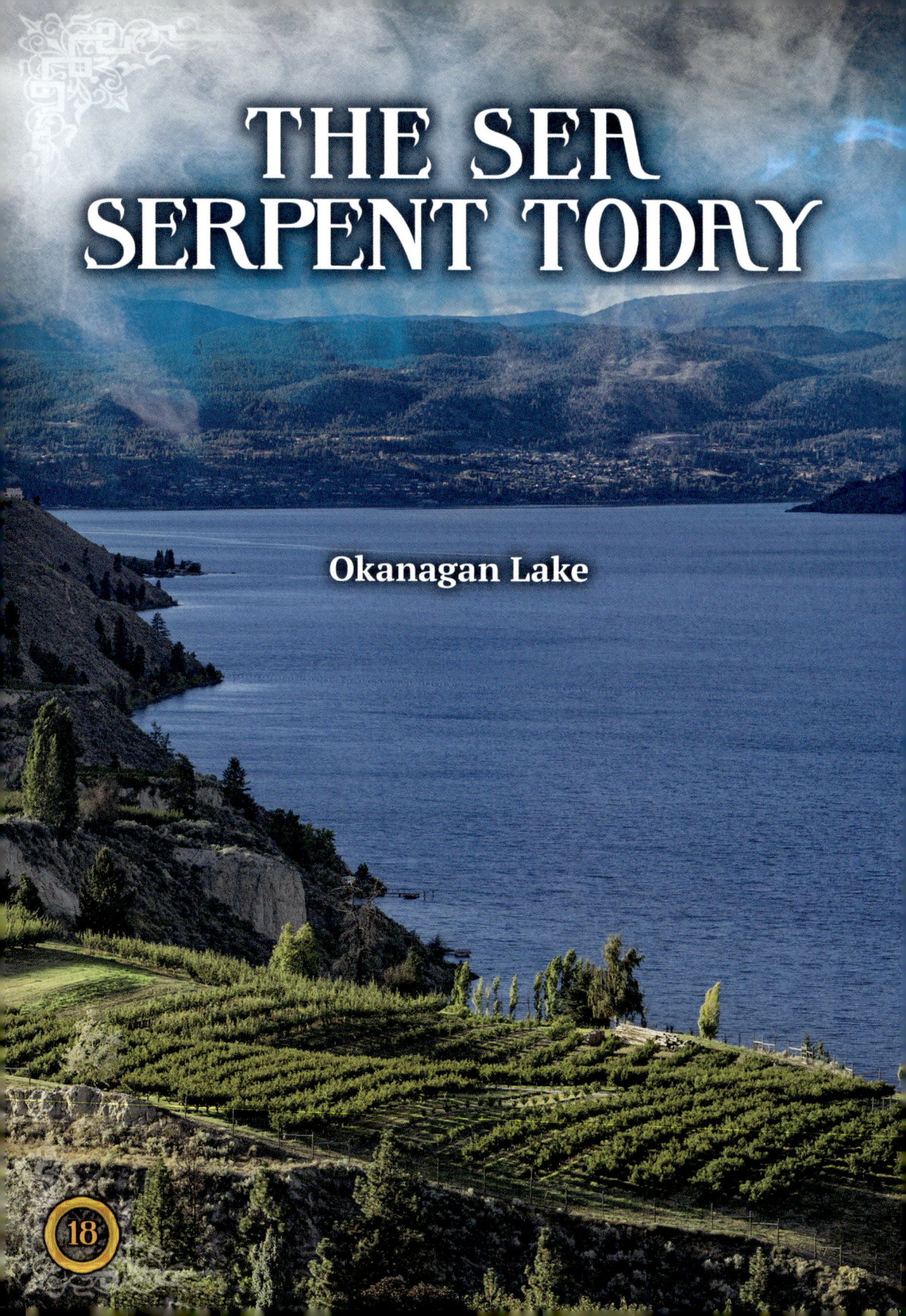

Sea serpents continue to be popular today. Ogopogo is a lake monster of Canadian **folklore**. It is believed to live in Okanagan Lake. Today, visitors try to spot the creature.

Sea serpents are used as symbols, too. A Norwegian **coat of arms** features a sea snake named Selma. In Norway, a company developed an **oceanic rocket** called Sea Serpent!

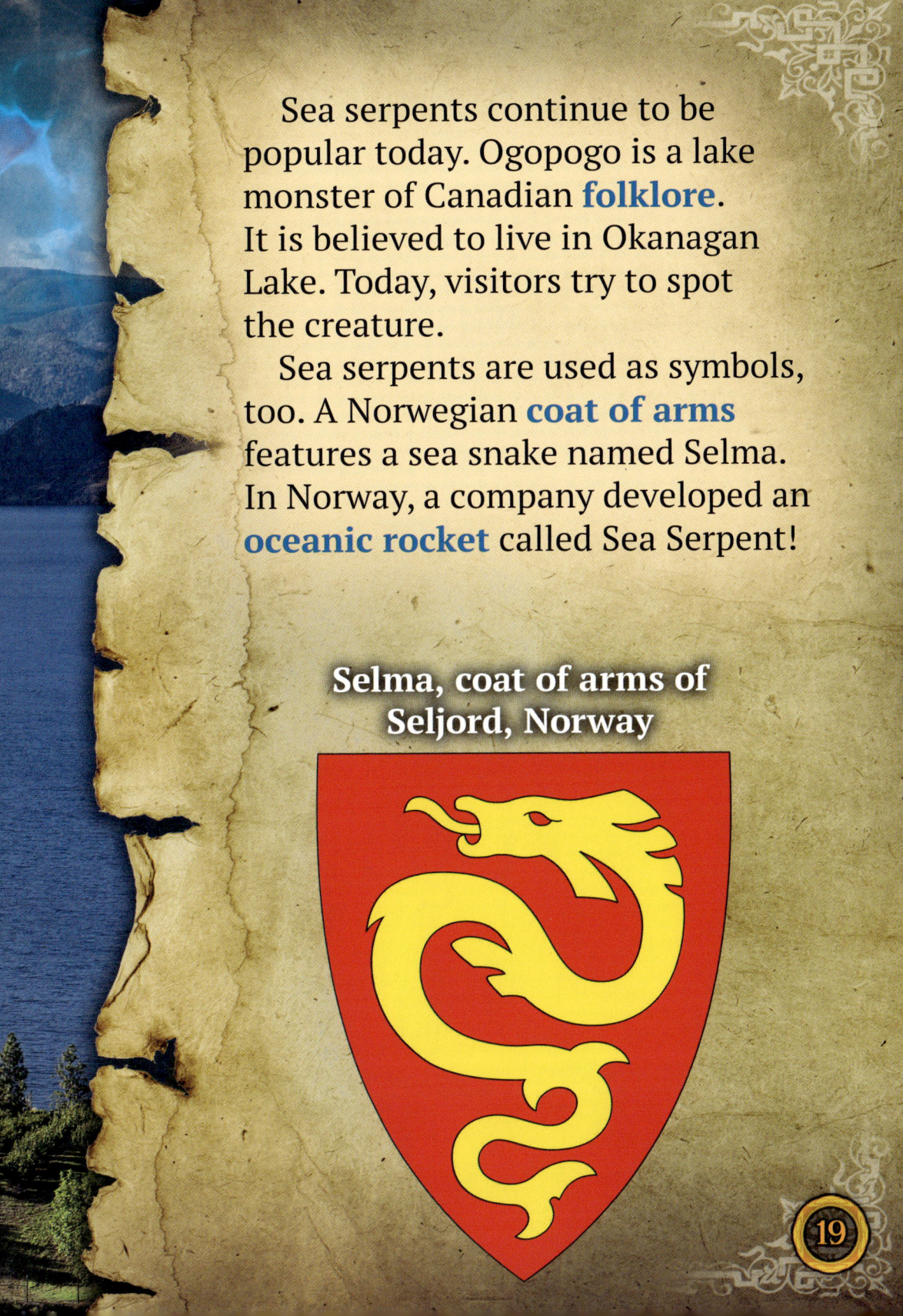

Selma, coat of arms of Seljord, Norway

Sea serpents show up in art, too! In 2019, the Jacksonville, Florida, winter beach festival Deck the Chairs featured a giant sea serpent sculpture. It was made out of lights and recycled plastic bottles.

The sea serpent still swims through people's imaginations. People around the world continue to wonder about this mythical creature!

Media Mention

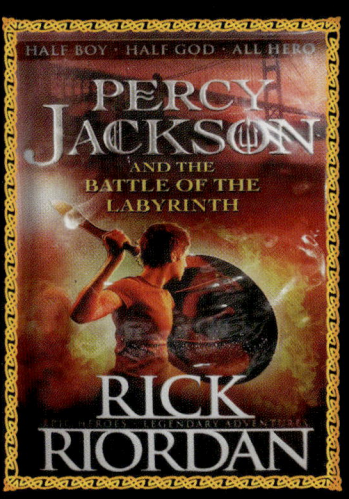

Series: Percy Jackson & The Olympians

Book: *The Battle of the Labyrinth*

Written By: Rick Riordan

Year Released: 2008

Summary: Percy watches people on a ship fight a giant sea serpent

Deck the Chairs sea serpent sculpture

GLOSSARY

chaos—a state of disorder

coat of arms—a special group of pictures or symbols that belong to a person, family, or group and are shown on a shield

creation myths—stories that tell how people believe the world began

cultures—the beliefs, values, and ways of life of a group of people

earthquakes—sudden movements of the earth's crust

folklore—the customs, beliefs, stories, and sayings of a group of people

Hebrew Bible—the holy text of the Jewish people; the Hebrew Bible also makes up part of the Christian Bible.

Mesopotamia—a region of southwestern Asia where many ancient civilizations began

mythology—ancient stories about the beliefs or history of a group of people; myths also try to explain events.

Norse mythology—ancient stories about the beliefs or history of the people of ancient Norway, Sweden, Denmark, and Iceland

oceanic rocket—a rocket that takes off from the ocean to go to space

symbol—something that stands for something else

tame—to bring from a wild to a gentle state

venomous—able to produce venom; venom is a kind of poison made by some snakes.

TO LEARN MORE

AT THE LIBRARY

Goddu, Krystyna Poray. *Sea Monsters: from Kraken to Nessie.* Minneapolis, Minn.: Lerner Publications, 2017.

Lawrence, Sandra, and Stuart Hill. *The Atlas of Monsters: Mythical Creatures from Around the World.* Philadelphia, Pa.: Running Press Kids, 2019.

Troupe, Thomas Kingsley. *Kraken.* Minneapolis, Minn.: Bellwether Media, 2021.

ON THE WEB

Factsurfer.com gives you a safe, fun way to find more information.

1. Go to www.factsurfer.com

2. Enter "sea serpents" into the search box and click 🔍.

3. Select your book cover to see a list of related content.

INDEX

Apep, 10
appearance, 7, 12
Aristotle, 11
around the world, 9
art, 20
Battle of the Labyrinth, The, 21
Chessie, 17
Chinese dragons, 7
Deck the Chairs festival, 20, 21
explanations, 16
folklore, 19
Hebrew Bible, 12
history, 10, 11, 12, 14, 17, 20
Jacksonville, Florida, 20
Jormungand, 14, 15
Leviathan, 12
maps, 8

Mesopotamia, 11
myths, 5, 7, 9, 10, 11, 14, 16, 20
Norway, 19
ocean, 7, 8
oceanic rocket, 19
Ogopogo, 19
Okanagan Lake, 18, 19
origin, 11
powers, 8, 10, 11, 12, 14
sailors, 5, 7, 8
Selma, 19
sightings, 16, 17
size, 4, 7, 12, 20
symbol, 11, 12, 19
Thor, 14, 15
Tiamat, 11
timeline, 12-13

The images in this book are reproduced through the courtesy of: Herschel Hoffmeyer, front cover (skin texture), p. 3 (figure); Michael Rosskothen, front cover (body); Cesare Sent, front cover (tongue); andrey polivanov, p. 3 (background); North Wind Picture Archives/ Alamy, pp. 4, 9 (bottom left); A.Dina, pp. 4-5 (serpent); m.mphoto, pp. 4-5 (background); DEA / Biblioteca Ambrosiana/ Getty, pp. 6-7; Narongdej Srithiyoth, p. 7; Olaus Magnus/ Wiki Commons, p. 8; Daniel Eskridge, p. 9 (top left); Science History Images/ Alamy pp. 9 (top right), 13 (top); Bejim p. 9 (bottom right); De Agostini / S. Vannini/ Newscom, pp. 10-11; Opera Nicolae/ Alamy, p. 12; Environmental Protection Agency/ Wiki Commons, p. 13 (bottom); Jef Thompson p. 14; Interfoto/ Alamy pp. 14-15; Jiang Hongyan, p. 16; Paulo Oliveira/ Alamy, pp. 16-17; Stan Jones, pp. 18-19; Bjarkan/ Wiki Commons, p. 19; sunthorn punaprung, p. 20; thepoo, pp. 20-21; Zety Akhzar, p. 21 (top); suneomp, p. 22.